W9-AZB-813

WHAT ARE MBUNA?

How many words in an African language do you know? Well, many tropical fish hobbyists use at least one on a regular basis—*mbuna*, actually a two-syllable word (ᵐbu-na) but pronounced by most English speakers as three: "um-BOO-na." In the local cherished by hobbyists the world over as beautiful and fascinating pets when kept alive and reproducing in aquaria.

Were you to undergo the hassles and hazards of traveling to Malawi, braving lakeflies and crocodiles, and dive into the

Photo by Mark Smith.

Many of the cichlids from Lake Malawi, the rockfish which act like freshwater reef fish, are extremely colorful, especially the sexually active males. This is a male *Copadichromis borleyi* from Eccles Reef in Lake Malawi.

language of the fishermen of the nation of Malawi the word *mbuna* means "rockfish" and refers to the numerous species of small rock-dwelling cichlid fishes that the people harvest to the tune of thousands of pounds per day as a major protein source in their diet. These same fishes, which are relished by these people as dinner when cooked or dried, are boulder-strewn coastal waters of Lake Malawi, the vast majority of the thousands of fishes you'd see would be cichlids, and most of these would be mbuna, many of them blue with black bars.

Of course, you don't have to do this to see mbuna; your local pet shop probably has quite a selection of them. It might also have a few other types of Malawi

Photo by Andre Roth.

Copadichromis   (Protomelas) chrysonotus **is not an mbuna but one of the open water species which has become popularized in the hobby because Dr. Axelrod brought it back to America in 1954 mistaking it for an mbuna when he introduced Malawi cichlids to the hobby.**

The local inhabitants of Lake Malawi depend upon the fishes for their protein. The fish are traditionally cooked over open fires. Millions of pounds of mbuna are eaten annually.

cichlids, most likely Malawi peacock cichlids or fishes of the type called *utaka* by the Malawi fishermen. The peacocks are beautiful, with impossibly brilliant blues, reds, and yellows; they need the same care as mbuna, but they do not mix well with them, since the peacocks are much less aggressive and will be oppressed by the overbearing mbuna.

The utaka live out over more open areas than the mbuna. Because of their large size and specialized behavior, not many of the utaka are suitable for the aquarium, but a few gorgeous smaller ones, such as the *Copadichromis* species, are hobby regulars. Aside from their preference for more open spaces,

Photo by Ad Konings.

Photo by Dr. Herbert R. Axelrod.

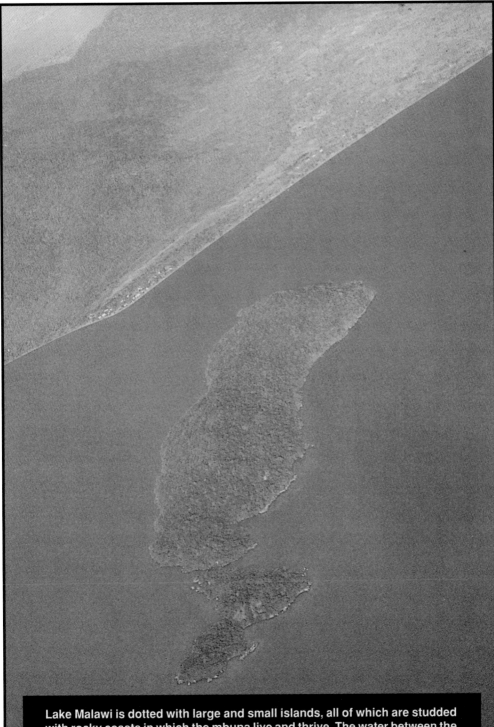

Lake Malawi is dotted with large and small islands, all of which are studded with rocky coasts in which the mbuna live and thrive. The water between the islands is deep and patrolled by predacious fishes which feed upon the mbuna, thus inhibiting contact between the mbuna of adjacent reefs. This resulted in the speciation of the mbunas. This is Thumbi Island West.

their care is much like that of the mbuna, so if you are attracted to these fishes, the husbandry techniques described in this book for the mbuna will serve you and your pets well. These two types can usually be housed together.

The use of the term *mbuna* is a bit variable, and you might hear a given species referred to both as mbuna and not. For example, I've seen *Protomelas* species referred to both ways, and while my *Protomelas* don't use the rockpiles the way my mbuna do, they're otherwise quite similar. After all, the cichlids of Lake Malawi are all descended from a few ancestral species, perhaps just one, so it is clear that like all of our other arbitrary divisions of nature, this one describes more a gradient than discrete categories. We hobbyists make up in enthusiasm what we might lack in linguistic precision, however (though I wonder whether the Malawi fishermen care very much about which species they're eating, as long as it tastes good). Fortunately, the care for all Malawi cichlids is similar; provided you select tankmates and diet correctly, it isn't that important whether you get the native names right for these gems of the African Rift.

If you choose to add a tank or two of these cichlids to your collection, you will certainly be rewarded with years of enjoyment from keeping and breeding the unusual and beautiful mbuna from Lake Malawi.

Photo by MP&C Piednoir Aqua Press.

Protomelas (Haplochromis) fenestratus.

LAKE MALAWI AND ITS FISHES

The cichlids of the African Rift lakes are among the wonders of nature. Formed in recent geological time by a rift, or separation in the earth's crust, these sea-sized lakes are unique among freshwater habitats; they are in many ways more similar to marine habitats. High mineral concentrations, stable temperature and pH values, reef structures (coral in the ocean, rock in the lakes), and even waves with surf zones along the shore are all shared by both the Rift lakes and tropical reefs. And the fishes of the two habitats are even similar in appearance. Perhaps not quite as dazzling as reef fishes, many mbuna are brightly colored or strikingly patterned, with bright blues, yellows, and reds, dark barring, and various contrastive color spots, stripes, or iridescence common in the color schemes. While the most spectacular marine tropicals surpass the prettiest mbuna, these cichlids remain among the most colorful of all fishes. They certainly provide aquarists with a diverse palette with which to decorate their aquaria and thus their homes.

Photo by Dr. Herbert R. Axelrod.

An aerial view of the Great Rift Valley which splits a great deal of Africa and was the basis for the formation of the Rift lakes, including Lakes Malawi, Tanganyika, Victoria, Kivu and Edward. Each of these lakes has its share of cichlids very few of which are not endemic (found only there).

Because of both their water requirements and their aggressive behavior, mbuna are not normally recommended for the typical "community" aquarium, but these fascinating fishes are ideally suited to a mixed Malawi collection, making many hobbyists who specialize in them into "biotype aquarists." If you are interested in joining their ranks, this book will provide you with the information you need to get started, plus an overview of available species.

MBUNA SPECIALIZATION

The Rift lakes and cichlids were made for each other. These fishes had several preadaptations that enabled them to quickly fill the various niches in this new ecology. Among them is their relatively recent marine ancestry, which put them in better stead than most other fishes for tolerating the saline chemical composition of the water. Malawi's water is saline in the sense of containing high concentrations of salts, though there is not a lot of sodium chloride, or table salt, which is, of course, the major salt in seawater. Chemically, however, especially as it concerns osmotic regulation, salts generally have similar biological effects.

Additionally, the cichlid trait of filling their swim bladders with gas generated within their bodies saves their fry from the danger most other fishes face of having to fill their swim bladder at the surface with atmospheric air, which, in the Rift lakes, would put them on the menu of just about every other fish in the lake.

Instead, mbuna fry scoot in and out of their mother's mouth, never having to rise up to the surface, and later hide in crevices in the rocks.

Likewise, the extended parental care typical of the family is an efficient response to the heavy predation pressures of the habitat of the lake. Mouthbrooding has developed in other cichlid groups as well, but in the Rift lakes it became the predominant reproductive strategy. Cichlids as a group opt for smaller spawns and greater parental care (even a large oscar's spawn of 2,000 eggs is paltry compared to, say, a cod's millions), but mouthbrooding enables the fish to produce only a few very large eggs. This results in much larger fry, which, of course,

Photo by W. Rose.

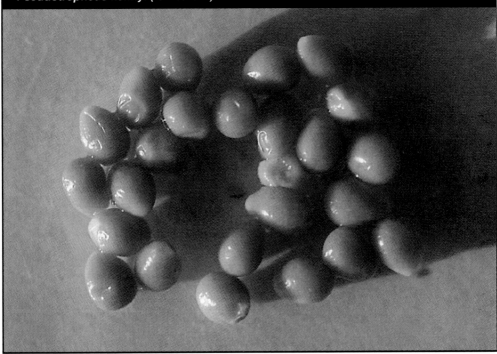

Typical mbuna eggs, carried in the parents' mouths for protection, are the basis for the success of the mbuna on an evolutionary basis. These eggs were carried by *Pseudotropheus kennyi (lombardoi).*

Photo by Dr. Herbert R. Axelrod.

Not all of Lake Malawi has rocky shallows and deep valleys. The shores are often lined with heavy growths of water plants in which many fishes live. No mbuna are found here, but many utaka (deepwater predatory species of cichlids) are found in their juvenile form.

means that they will be able to fend for themselves that much sooner. While they are most vulnerable they reside in their mother's buccal cavity, actually more her throat than her mouth.

The physical habitat of the lake is a bit austere or, at least, uncomplicated. There is some higher plant life, but there is an even greater abundance of algae, fueled by a tropical sun that is largely unhindered by murkiness or turbidity in the water, as Malawi boasts some of the clearest fresh water in the world. Though there are sandy shores, mbuna are associated with the rocky areas, which extend from the shore right down into the bottom of the lake. From gigantic boulders to small stones, this jumble of a rockpile provides both myriad nooks and crannies for hiding places and copious surface area for algal growth.

The algae, in turn, provide dense foraging for various invertebrates. This has provided aquarists with another foreign word, *aufwuchs*. This German term refers to the salad and seafood smorgasbord growing on the rocks, upon which so many mbuna graze.

While food for the fishes is fairly plentiful, competition is fierce, and the principal food sources are few: plankton (microorganisms in the water column), aufwuchs (algae and invertebrates), and other fishes (mostly cichlids). There are specialized piscivores, both the large ambush predators

Nimbochromis (Haplochromis) livingstoni is a widely available utaka which is predatory on smaller fishes. They wear a coat of camouflage. Photo by Dr. Herbert R. Axelrod.

like *Nimbochromis* and fry eaters like some *Melanochromis*, but no mbuna is going to pass up a meal "on the fin" if something small enough swims close enough. There is a strong drive among Malawi cichlids to find protein, and other fishes are, of course, a wonderful source of protein.

This, however, is true for almost any fishes anywhere. What makes the Rift lake situation different is that the fertile habitable zone of the lake is relatively a very small area. The lower and unoxygenated stratum is, of course, a barrier, but most species are confined to an even smaller zone, specifically the numerous rocky reefs and shorelines that punctuate the exposed sandy or marshy areas and wide, barren stretches of open water. The effect is of fertile islands, bustling with life, amid large expanses relatively devoid of life. This means that the large populations of fishes live at very high densities, very similar to the frenetic bustle of a human urban community. This high concentration of resources and consumers, unlike the more homogeneous distribution of most other freshwater habitats, has been a major influence in the evolution of mbuna.

EVOLUTION IN ACTION

The incredibly rapid speciation in the Rift lakes is a biologist's dream; nowhere else has evolution so obviously and swiftly taken place. Darwin's famous Galapagos finches are but a

Photo by Dr. Herbert R. Axelrod.

Dr. Axelrod brought this fish back as a one-inch juvenile he caught in the grassy areas of Lake Malawi. It grew into a 10-inch predator which, because it bred readily, was available to aquarists in the early 1950's. This is known as *Nimbochromis (Haplochromis) fuscotaeniatus.*

Photo by Dr. Herbert R. Axelrod.

Many mbuna species exist in several color varieties. These varieties of the same species are called *morphs* with variations in body color, fin edging colors and striping. This is a *Melanochromis* species among the first brought back by Dr. Axelrod in 1952.

footnote compared to the drama of the biota in these lakes. In a few thousands of years (an eyeblink in a geologic or evolutionary time scale) hundreds and hundreds of new species have arisen from a handful of original species. This, however, produces a nightmarish situation for the taxonomist; the problem stems from having so many similar species, many with various color morphs, some with pronounced sexual dimorphism—and of the latter, many with only seasonal dimorphism, with males assuming breeding dress during only part of the year. All of this results from the type of habitat Lake Malawi provides.

A BIOLOGICAL NOTE

While it is not uncommon for some animals (notably birds and fishes) to be able to produce viable and sometimes fertile hybrids, nowhere is this more prevalent than among the mbuna. This makes sense if you consider the speeded-up rate of evolution in the Rift lakes. Although many tend to think of a given species as the endpoint of an evolutionary line, all species are, of course, transitional; it's just that human beings haven't been around long enough to notice much change.

With mbuna, we see the transitional nature of species much more clearly. In the lake, a very tight balance exists among various selective pressures. For example, as certain types of males get better and better at attracting females, varieties drift toward becoming separate species, and species drift even further apart. When a given female starts to be attracted by a "wrong" male, there will normally be a "correct" male nearby to compete for and win her attentions.

In the aquarium, however, with the fish permanently locked into the same tiny area, the chance of a female's succumbing to the display of a given male is highly increased. Should there be no correct male, or should she find the single present male of her own species less attractive for any reason (including that he is off spawning with another female), she will mate with another male, and because of the close

relatedness among the species, the chances are the mating will produce viable young.

If you watch a tankful of mbuna, you will see that when a male begins to display to a female many other males get into the act. Often things work as nature intended, and she ignores all the males of other species. Sometimes, however, she does not. Thus, while males in the lake are competing against all the other males of their species and the female gets to pick her favorite, in the aquarium there is much more competition *across* species, since we usually keep only one or two males of a given species per tank. Under those circumstances, the female's favorite could easily be a male of another species.

Captive hybridization is undoubtedly inevitable, but

Photo by Dr. Herbert R. Axelrod.

These are two *Pseudotropheus tropheops*. Note the difference in coloration of their dorsal fins. This is one type of variation found within the same species.

Photo by Andre Roth.

Two color morphs of *Pseudotropheus zebra*.

responsible hobbyists will try to minimize such accidents and will never set up a hybrid mating on purpose. If hybrids are produced, they should either be used as feeders or kept just for the enjoyment of their beauty, without saving any progeny from them. Giving them away is not a good choice, since from that point you have no control over how they are used in breeding operations.

MBUNA BY ANY OTHER NAME

Mbuna taxonomy has come a long way from when almost every fish was classified as "*Haplochromis*," but as soon as one group seems to be sorted out, some new form is exported and the confusion starts all over again. Add to this the propensity of these species to hybridize in captivity (and to a small extent in the lake itself) and you can see the problems inherent here.

These problems will affect you as a hobbyist in trying to obtain purebred stock without paying the extremely high fees for freshly imported fish, and in trying to identify the fish you see for sale. Just because a distributor or a dealer sticks a scientific name on a tag on the tank doesn't mean it is a correct taxonomic identification. And any mistakes are probably not the dealer's fault, since the best way to identify some species is to be sure of their source.

Why? Mbuna species are, as we've said, highly variable, often containing dozens of morphs, all of which freely interbreed. In addition, many of these morphs look the same from species to species, so, for example, many species have a dark blue male with dark bars, an OB (orange-blotched) female, a light blue male with dark bars and red dorsal edge, a solid red/orange morph, a creamy blue morph, a solid blue morph, etc.

The way to be certain is to learn as much as you can about mbuna taxonomy and to carefully choose a reputable dealer who will not offer vaguely identified fish just to provide a bargain price. It costs something to keep track of a fish each step of the way from its mother's mouth to a dealer's tanks, but it is usually worth the price.

If all you want is some pretty fish for a tank or two, it obviously does not matter whether they are

purebred or whether you know the correct scientific name for them. Most hobbyists, however, are committed to scientific accuracy and the perpetuation of pure lines, and hobbyists contemplating breeding their fishes should be careful about their identification. In no case should you be tempted by a sign offering "mixed African cichlids" at a cheap price *unless* you are just looking for a beautiful display of fish.

Because of the incredibly large potential variation within a species of mbuna and the fact that several genera have at least one species looking like those of another genus, even ichthyologists would have trouble identifying a jumble of mbuna, so there is no way that you are going to be able to raise up a mixed batch and then pick out appropriate fish to place together for breeding just by their appearance.

A well planned mbuna tank featuring *Melanochromis auratus.* The two darker fish are sexually active males.

WHICH MBUNA SHOULD I TRY?

Typically books about various types of tropical fishes begin with a discussion of their care, reserving a description of the species themselves for later on. Besides the fact that I'm eager to talk about these gorgeous fishes, I'm going to discuss a selection of mbuna species up front so that you can be thinking about the ones that appeal to you most as you read about their needs and how you can best provide for them. It is a great temptation to go into a pet shop and buy a few of these and a few of those, but you will be most successful with mbuna if you populate your aquarium first in your head or on paper. Problems such as predation, aggressive incompatibility, and hybridization can largely be avoided with proper planning.

While the following descriptions are of necessity brief, and while I can include but a fraction of the species of mbuna, even if we restrict the discussion to those usually available in the hobby, and while photos, marvelous as they are, cannot do complete

Photo by Andre Roth.

A pair of *Melanochromis simulans*; the male is the darker fish in his spawning dress.

Photo by MP&C Piednoir Aqu

Less aggressive than many other Mbuna, *Labidochromis caeruleus* is usually seen in this coloration, although other color morphs, the eponymous blue, for instance, are occasionally seen.

justice to the way the vibrant colors of these fishes play against the light and shadows in the aquarium, this section should give you a good idea of what you have to choose from— it will certainly help you to recognize and identify most of the mbuna you see at your local dealer.

It is particularly important to consult a good atlas that has color photos of adult cichlids before deciding what to buy. Usually the most spectacular colorations are found on breeding males, and the fish commonly available in stores are in the drab, immature, non-breeding and female colorations. This is particularly noticeable with the utaka and the peacocks, which are typically nondescript silver fish until the males explode into their nuptial dress. Your dealer should have a reference book in which he or she can show you adult pictures of the fish you are interested in.

There are many books on cichlids, but the most complete and most well illustrated is the *Lexicon of Cichlids*, TS-190, available from any aquarium shop. Even if you don't buy it, at least look through it and enter the dazzling world of intelligent fishes known as cichlids.

PSEUDOTROPHEUS

You cannot go far in this area of our hobby without encountering this name. Literally it means

"false Tropheus," but practically it means serious headaches for taxonomists and potential confusion for aquarists (who are often one and the same). This problem is not unique to this genus among Rift lake cichlids, even if it is perhaps most pronounced with it.

It sometimes seems as if the fishes of the genus *Pseudotropheus* took their biological preadaptation to extremes, even for Rift lake cichlids. One of the pressures on these populations was species self-identification for mating. When polygamous males defend territories and attract females to them, selection works to enhance any trait that helps females to identify the males of their own kind. This creates a wide variety of morphs. Among *Pseudotropheus* there are many species that have numerous color morphs, often with similar females and very different males. In an atlas where most cichlid species have a photo or two, these fish will have pages and pages of photos, each one different.

In addition, several "species" are actually "complexes" of

Photo by Dr. Herbert R. Axelrod.

Dr. Axelrod donated a batch of mixed Malawi cichlids to the Steinhart Aquarium in San Francisco, California so they could grow to their full potential and their behavior could be studied. This photo, taken in 1979, shows the fishes many years after they were received by the aquarium. They behave the same way in a large aquarium as they do in Lake Malawi.

species where, in taxonomic desperation, similar types of fishes are lumped together without clear distinctions among them.

Pseudotropheus zebra

The epitome of the situation just described is found with this species-complex, an original mbuna in the trade. Decades before the explosion of Rift lake cichlid popularity, this fish made a tremendous splash in the hobby with its unusual habits, multiple color morphs, and exotic origin. It seemed that every shipment of the "Nyassa Cichlid" brought another color morph, and it was hard to believe that these bright blue, orange, striped, and mottled fishes could all be the same species. And they were mouthbrooders!

Because of their long tenure in the hobby, these fish are quite inexpensive, and also quite hybridized, at least among the various color morphs. There are also albino strains.

Of about average aggression for mbuna (which is a lot!), these fish are extremely easy to care for and breed. Most species are in the five-inch range, and they usually have marked sexual dichromatism, though there are morphs shared by both sexes. They are quite omnivorous, and always hungry!

Photo by Dr. Herbert R. Axelrod.

Pseudotropheus zebra color morph naturally appearing in Lake Malawi. This morph has hardly any stripes.

Photo by Dr. Herbert R. Axelrod.

One of the original *Pseudotropheus zebra* collected by Dr. Axelrod in 1954.

This is also a *Pseudotropheus zebra* collected by Dr. Axelrod in 1954. It is a so-called OB (=orange blotched) color morph.

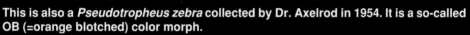

Photo by Dr. Herbert R. Axelrod.

Photo by Dr. Karl Knaack.

Another color morph of *Pseudotropheus zebra*.

Still another color morph of *Pseudotropheus zebra* collected by Dr. Axelrod in Lake Malawi in 1954. It is important to have photos of these original fishes because most Malawi cichlids are now domestically bred. To further complicate matters, many of the islands in Lake Malawi were poisoned to kill all of the natural populations, and more desirable (for aquarium purposes) fishes were introduced. Since the poison didn't kill all the fishes, a great number of hybrids developed, many of which were subsequently described as new species.

Photo by Dr. Herbert R. Axelrod.

They will hybridize freely among the species of the genus, and even with some species in other genera. If you are serious about breeding these fishes, you should maintain only one *P. zebra* morph per tank. It is usually best to have one male and several females per tank as well.

Pseudotropheus sp. "Acei"

This beautiful fish is quite plentiful in the hobby. It is an unbarred mbuna, usually a brownish purple or blue, often with contrasting white or yellow tail. There is not much difference between the sexes.

Photo by Dr. Herbert R. Axelrod.

Pseudotropheus 'acei' is really a morph of *Pseudotropheus torsiops* as proven by the 1954 photo and specimen collected by Dr. Axelrod.

Pseudotropheus kennyi (lombardoi), a dominant male (sexually active).

Photo by Adam Kapralski.

A female *Pseudotropheus kennyi (lombardoi)* with a mouthful of eggs.

Photo by Burkhard Kahl.

Like all other *Pseudotropheus*, they are easy to keep and are reliable breeders.

Pseudotropheus kennyi (lombardoi)

A commonly available and inexpensive species, this zebra-barred cichlid is unusual in that it is the males which are yellow and the females blue, the reversal of typical mbuna blue/yellow sexual dichromatism. A truly beautiful fish, it has the unfortunate trait of being even more pugnacious and aggressive than usual for mbuna. Some hobbyists are successful keeping it in a mixed community, and others experience terrible mortality from their bullying and territoriality. I've had mixed

Photo by Dr. Harry Grier.

A cultivated albino *Pseudotropheus kennyi (lombardoi)* produced in Florida and a prize-winner at the Florida Tropical Fish Farmers Association fish competition.

Photo by Adam Kapralski.

success combining *P. kennyi* with other equally large mbuna. In any case, it is difficult to support two breeding males of this species in one tank, even if it is extremely large. They are ready breeders with good-sized clutches, and the fry, like all other mbuna babies, are easy to raise.

Pseudotropheus aurora

A particularly striking *Pseudotropheus*, with muddy brown females but light blue and yellow males.

Pseudotropheus tropheops

This name covers another complex of species with extremely variable forms. Even if you cannot

Pseudotropheus aurora male chewing on a snail. Males typically are yellow and light blue.

The muddy color of a female Pseudotropheus aurora.

Photo by Adam Kapralski.

identify which particular strain you have, you can at least keep different morphs separate for breeding. It is likely that by the time taxonomists come to any agreement about this species-complex all the aquarium specimens representing it will have beeen hopelessly hybridized. We should try to stem the tide, however, if possible.

Pseudotropheus saulosi

One of the smaller species of the genus, it matures at about three inches. It is commonly seen in a solid orange-yellow morph, one of the less common mbuna forms. This morph is easily confused with the "red zebra," which you likely will see labeled a *P. zebra* morph even though it is now considered a separate species. There is no difference between the sexes in coloration.

Pseudotropheus "socolofi"

A very popular aquarium fish, this "species" has a dubious validity. It is apparently the result of the hybridization in Lake Malawi of species that had been transplanted to non-native reefs by people wanting to have a closer source of fish to catch for export.

Two color morphs of *Pseudotropheus saulosi*. There are no apparent sex differences.

The so-called *Pseudotropheus* socolofi is not a natural species. It was produced by unnatural hybrids in Lake Malawi when Peter Davies poisoned a reef and transplanted more desirable species onto that reef. This fish is popular and appears in an albino form.

Photo by Mark Smith.

A young *Pseudotropheus crabro*, the so-called bumble bee cichlid.

Its origin aside, it is a beautiful purplish-blue fish, with no marked sexual dimorphism. It is available as well in an albino strain. I once had and enjoyed a strikingly colorful aquarium containing equal numbers of *P.* "socolofi" and red-blotch zebras.

Pseudotropheus crabro

Here is an odd mbuna, odd in that it shows very little variation and no real geographical variants, and it is not easily confused with any other species. Amazing! Often sold as the "bumble bee" cichlid, it has alternating yellow and dark bars. This unique coloration makes it a good contrast in a mixed mbuna community.

Photo by Adam Kapralski.

A mature *Pseudotropheus crabro,* the bumble bee cichlid. Photo by MP&C Piednoir Aqua Press.

pheus sprengerae

This is just about the perfect beginner's mbuna. Subtly beautiful with its rich red-brown base color and overwashes that can range from gold to purple, very peaceful for a mbuna, hardy, and easy to breed, the common "rusty cichlid" is a friend to anyone who's just getting started with Malawi cichlids.

In fact, these fish are so peaceful for mbunas that they should not be kept with most other species, since they will be harassed and perhaps even killed by the more aggressive fishes. This is unfortunate, since their coloration is atypical for mbuna and they make a nice color contrast. You may be successful in keeping them in a mixed collection, with the help of careful choice of companions and a bit of luck, especially if you choose some of the milder mbuna, like *Labidichromis*, or some peacocks. In this latter case, it may be the *Iodotropheus* who wind up being too aggressive.

These fish breed like crazy, and you will probably soon tire of netting out holding females and start letting the fry fend for themselves in the parents' tank instead, which they often do successfully. This mbuna needs a lot of vegetable matter in its diet, and my rusty tanks never have any algae buildup.

Photo by Jaroslav Elias.

Iodotropheus sprengerae, the rusty cichlid, is an ideal fish for the beginner.

Iodotropheus sprengerae, a male from Boadzulu Island. this species seems to favor a single egg spot in the anal fin.

Photo by Ad Konings.

Melanochromis

This genus has several aquarium favorites. Most of them have marked sexual dichromatism

Melanochromis johanni, a male in dominant dress.

of the seasonal type; i.e., in any population of like-colored females, young, and nonbreeding males, there are also very differently colored, sexually active, territory-defending males.

A tankful of these fishes might have only one or two visible males in it (rarely more due to intense male aggression), but if they are removed, very soon some of the inactive males will color up and stake out breeding territories.

In the case of *M. auratus*, the difference is especially amazing, since the male coloration is the

Photo by Heinrich Stolz.

exact opposite of the female-juvenile pattern: gold and neon blue stripes on a black background versus black and neon blue stripes on a gold background. The change in color happens very quickly, with the fish looking like a double exposure of both patterns for a couple of days in between. This is a truly rough character, and I have had a single male dominate an entire tank of cichlids in a reign of terror.

Another favorite is *M. johanni*, often called the "electric blue johanni." And electric they are! The iridescent blue on black of a breeding male of this species is spectacular. Females range from muddy brown to washed-out blue and black to yellow. I breed two strains; in one the inactive and female coloration is blue on silver, with the active male coloration being blue and black, and in the other, which is considerably smaller, the default coloration is a solid yellowish orange, while the active male morph is bright purple and black, making the transitional males come out a strange muddy purple. The first time I saw one changing coloration, I thought it was dying of some exotic disease! Of course, in a few days the metamorphosis is complete, and a tankful of these fish is a spectacular sight, holding as it does a group of dark electric blue males along with neon-bright orange females and inactive males.

Photo by Dr. Herbert R. Axelrod.

An inactive male *Melanochromis johanni* photographed by Dr. Axelrod at Lake Malawi. The species was named to honor one of the native fishermen who drowned in the lake while fishing for mbuna.

Photo by Edward Taylor.

Melanochromis johanni.

Melanochromis auratus collected in Lake Malawi by Dr. Axelrod.

Photo by Dr. Herbert R. Axelrod.

The original dominant male *Melanochromis johanni* collected by Dr. Axelrod in Lake Malawi. Photo by Dr. Herbert R. Axelrod.

One of the numerous color morphs of *Cynotilapia afra*.

Photo by Gerald Meola, African Fish Imports.

Cynotilapia afra

This is basically a smaller version of a *Pseudotropheus*-like cichlid. It, of course, comes in various morphs, and many of them differ in the color of the band at the top of the male's dorsal fin. The band can range from white to yellow to red to black.

While not a "dwarf," this fish is one of the smaller mbuna. My personal opinion is that this is the prettiest of all mbuna that are blue with black bars; there is something especially brilliant about its coloration, with a velvet richness quite similar to that of *Melanochromis johanni*.

Labidichromis

Though scientifically named for its sky-blue morph, *L. caeruleus* is

Cynotilapia afra, the blue morph.

Photo by MP&C Piednoir Aqua Press.

Photo by Mark Smith.

Cynotilapia afra, a lovely color morph.

Labidochromis caeruleus. The yellow morph looks nothing like the blue form for which the species was named.

a favorite in its yellow form. Sold as the "lemon drop cichlid," this relatively placid bright yellow fish is a welcome change from the blues, blacks, oranges, and purples that predominate in mbuna color schemes. The black edge of the dorsal and anal fins contrasts nicely with the overall coloration. While some rely on the presence or intensity of this black to indicate male sex, I have had known females (that is, with broods in their mouths) that had heavy dark bands. As with many other mbuna, males tend to be a bit larger and heftier; after all, they never have to undergo two-week fasts while they incubate eggs.

L. chisumulae is a particularly pleasant fish, also known as the

Labidochromis caeruleus collected and photographed at Lake Malawi.

Photo by Dr. Warren E. Burgess.

A **Labeotropheus** of unknown genetic origin but produced in an albino form. It is sold as **Labeotropheus trewavasae.**

Photo by Dr. Herbert R. Axelrod.

Labeotropheus trewavasae, the red stripe cichlid.

Photo by Mark Smith.

Likoma clown. It is diminutive, nicely colored, and fairly peaceful. The female coloration is an extremely pale blue, and the male superimposes a pattern of dark bars and a dark throat.

Labeotropheus

Are there one or two species? The difference between L. trewavasae and L. fuelleborni is basically body shape, with the former being more slender. Both show a wide variety of color morphs, including the typical blue with black bars, and ranging to orange, pink, or pale gray, either solid or mottled with black, blue, or orange. While these colorations are similar to those of many other mbuna, these fishes are easily distinguished by their downturned snout.

Photo by Dr. Herbert R. Axelrod.

Here is one mbuna for which you absolutely must not scrimp on the vegetable matter in the diet. This fish is truly adapted to grazing algae, with a mouth positioned so that it can scrape rocks without standing on its nose. In the artificial environment of an aquarium, this otherwise useful feature causes them to have to approach scraping the glass walls at odd angles, and they look as maladapted doing this as other fishes do when trying to graze horizontal surfaces.

Protomelas

Several species of this genus are available in the hobby, including *P. taeniolatus*, often

Labeotropheus fuelleborni (?), the OB (=orange blotched) color morph. This is a natural form collected in Lake Malawi.

Protomelas (Haplochromis) fenestratus is a lovely fish.

Photo by Adam Kapralski.

Protomelas (Haplochromis) taeniolatus, the red empress cichlid.

Photo by MP&C Piednoir Aqua Press.

called the "red empress." All of these species are highly variable, with every reef in the lake yielding a different morph and every hatchery in the trade supplying its own version of *P. taeniolatus.*

Even photos hardly do these fishes justice, and although the juveniles you find at your dealer's may not look like much, remember what the adult males look like. You won't be disappointed if you give them a try.

Nimbochromis

The several species of this genus are personal favorites of mine, but they make you pay for enjoying them. They are not usually classified as mbuna, but their diet does consist largely of mbuna. These are BIG fishes, eight to ten inches or more. And scrappy. And voracious. I have had considerable difficulty establishing a stable community, even in a 100-gallon tank. Fights to the death were common, but oh, are they beauties!

First of all, the basic camouflage color of dark brown splotches on a light background is attractively distinctive. Then the breeding male colorations are breathtaking, with blues predominating, along with yellow accents. These males rival coral reef beauties when in their intense nuptial coloration. Breeding males may be difficult to distinguish, but males of the commonly available species are fairly easily differentiated in nonbreeding coloration.

Nimbochromis (Haplochromis) venustus.

Photo by MP&C Piednoir Aqua Press.

N. venustus has a yellow wash over the body, *N. livingstoni* does not, and *N. polystigma* has a freckling of tiny dots over the body and fins in addition to the camouflage pattern. When available, *N. linni* is distinguishable by its downturned

Nimbochromis (Haplochromis) livingstoni, the predatory utaka. Fish collected by Dr. Axelrod with hook and line in Lake Malawi.

Photo by Dr. Herbert R. Axelrod.

Nimbochromis (Haplochromis) linni known for its unique tapir-like snout.

Photo by Edward Taylor.

tapir-like snout. There is probably taxonomic revision yet to be made here, but a fifth species *N. fuscotaeniatus*, is normally recognized.

The coloration of these fishes truly is a camouflage tactic; they will actually lie on the bottom of the lake as if dead, blending into the sand and rocks while waiting in ambush for unsuspecting prey. Sometimes one will do this outside a rock crevice housing a group of fry, jumping to life when curious little fish swim too far away from the shelter, sucking them all into its cavernous mouth.

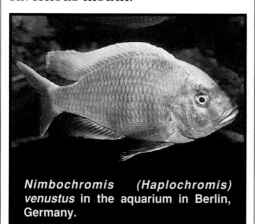

Nimbochromis (Haplochromis) venustus in the aquarium in Berlin, Germany.

Photo by Dr. Herbert R. Axelrod.

Provided you can raise a group to maturity, they spawn like typical mouthbrooders, and the females carry the young for a long time, a month or more.

Copadichromis
The utaka of this genus are spectacularly beautiful in the male breeding coloration. Commonly available are *C. azureus*, obviously a brilliant,

Copadichromis (Haplochromis) azureus.

Photo by Mitsuyoshi Tatematsu courtesy of Midori Shobo.

metallic blue, *C. borleyi*, a highly variable species but especially striking in the morph with the deep crimson coloration, and *C. chrysonotus*, a less variable fish with a deep blue breeding male.

Aulonocara
These are the Malawi peacocks, a deserved name. You will see fish labeled *A. stuartgranti, A. baenschi, A. hansbaenschi, A. korneliae, A. maylandi*, and many others, with enormous variation within a "species." There is a riotous variation of colors in these fishes, and they simply have to be seen to be believed. Many sport

Aulonocara baenschi.

Photo by MP&C Piednoir Aqua Press.

contrasting "egg spots" that enhance their brilliance.

As was already mentioned, these less pugnacious Malawi cichlids do not fare well with mbuna, but a breeding group of one or two males and a harem of females are a sight well worth its own tank. In the lake they eat invertebrates that they locate in the sand, and they are definitely one of the more carnivorous groups of Malawi's cichlids.

Besides the extremely difficult taxonomic situation in nature, many captive fish are hybridized, and with all of the natural variation within the species, it is impossible to be certain. Trying to be part of the solution and not of the problem, keep your strains separate and don't crossbreed.

This extremely abbreviated survey of Malawi cichlids, plus the pictures in this book, should get you started recognizing and identifying the more commonly available species. If your dealer has something else, check it out in a good reference book, and if it appeals to you, enjoy!

Aulonocara maylandi.

Photo by Edward Taylor.

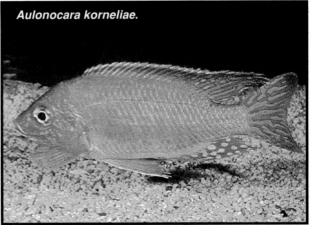

Aulonocara korneliae.

Photo by Kunihiko Ohki courtesy of Midori Shobo.

SETTING UP THE TANK

THE IMPORTANCE OF WATER

In order to prepare a tank for the mbuna you wish to obtain, you need to understand their requirements. The needs of these fishes are not complicated, though they are different from those of many of the fishes you might already be familiar with, even from those of other cichlids.

Specifically, they need very clean, hard, alkaline water. If you are lucky enough to have hard, alkaline water, you're all set. If your water does not test out at least moderately hard and with a pH of at least 7.5, preferably higher, you do not have to forgo the pleasures of keeping mbuna, but you will have to doctor their water. It is, fortunately, an easy matter to add minerals to water for these cichlids, unlike the task facing someone who has hard water and wishes to raise blackwater species like discus.

Prepackaged Malawi salt mixes are available which will provide buffered water of the proper pH and hardness. You must, of course, add these whenever you change water, which should be frequently. This must be accompanied by testing the hardness and pH, since you cannot otherwise know of differences due to water supply variance, uneven evaporation rates, biological processes, and the contribution of soluble rocks in the tank.

That's right, those substrates like crushed coral or dolomite and those coral and limestone rocks that you have been told must not be used in freshwater tanks are suitable for Rift lake tanks as well as for saltwater ones. Their buffering capability is insufficient to make unsuitable water fit for your mbuna, but as they dissolve they will raise pH and carbonate

Water conditions are crucial when maintaining a natural environment for Malawi cichlids. If your water isn't hard enough (few city waters are), you need a water conditioner. Photo courtesy of Mardel Laboratories.

hardness and serve to counteract acidic conditions that arise from biological activity.

This biological activity, including both fish metabolism and waste decomposition, does a lot more than just acidify the water. The principal waste product of concern is ammonia, which is a normal byproduct of your fishes' metabolism and which, like most other waste products, is highly toxic to them. In an aquarium with a properly functioning biological filter, the ammonia is taken up by bacteria that convert it to nitrites, which are slightly less toxic to the fish. Then other bacteria take up the nitrites and convert them to nitrates, which are substantially less toxic. You are probably used to monitoring nitrates and keeping their levels down with proper husbandry, including those frequent partial water changes. You have to be even more diligent with your mbuna aquarium, since these fishes, adapted to the pristine waters of Lake Malawi, are less tolerant of these pollutants than many fishes from less pure habitats.

Aquarium stores carry excellent salts for the proper maintenance of Malawi cichlids. Be sure to get a salt made especially for Malawi cichlids. *Don't use sea salts*! A number of manufacturers have developed a salt duplicating the salts from Lake Malawi. Photo courtesy of Tropic Marin.

THE AQUARIUM

Mbuna are mostly large, nasty, active cichlids. They need lots of room. For a community aquarium, a four-foot aquarium is about the minimum size, with the wider 70- or 90-gallon tank being preferable to the narrow 55-, since bottom area is important for spacing out the fishes' territories.

If you're setting up a single-species, single-male breeding tank, you may be able to use something smaller, but bigger is still better.

Because of their beautiful colors and lively manner, mbuna are perfect for large display aquaria, and a well chosen group of these cichlids in a 135-gallon or larger tank is a very impressive sight.

Also impressive, of course, is the *weight* of the mbuna tank, so the location and stand need to be carefully considered. Make sure the floor in the spot you have chosen can support the total weight, which you can estimate at ten to twelve pounds for each gallon of tank capacity. A solid commercial stand is necessary unless you can get a professionally designed and built homemade one.

FILTRATION

What about that biofilter? You may be used to using an undergravel filter (UGF) as a biofilter. The principal objection to a UGF in a mbuna tank results from the digging habits of these fishes. While they seek out a ready-made hole in the rocks, they will rearrange things to suit themselves, and some are quite proficient excavators. I have seen a single 2 $1/_2$-inch male mbuna move about five pounds of gravel out from under a rock in under an hour. This objection can be overcome by installing some sort of barrier—plastic screening or "egg crate" light

There are products available at your local pet shop which will make water changes easier and less messy. Photo courtesy of Aquarium Products Company.

diffuser— over a layer of gravel above the filter plate, topped with another layer of gravel in which the cichlids can dig to their hearts' content without disrupting the filtration.

Another objection, the most common against this somewhat dated workhorse of the hobby, is that the gravel and the plate can become clogged with debris. In a tankful of large-bodied, heavy-

feeding cichlids this can happen rather quickly. Coupling siphon vacuuming of the gravel with the regular water changes will go a long way to minimizing this problem, as can occasional vacuuming under the plate by slipping the siphon down a lift tube. Or else run a reverse-flow UGF system.

Many mbuna keepers rely on other biological filtration such as the numerous variations on the wet-dry theme — trickle filters or biowheel filters — or fluidized bed filters, which maximize surface area and water flow rather than oxygenation (a "wet-wet" filter, perhaps?). If you are familiar and comfortable with denitrators, either bacterial or electronic, they can, of course, be used on mbuna tanks as effectively as they are on saltwater or other freshwater tanks.

Both power filters and canister filters can be used for mechanical and chemical filtration in mbuna setups, though the typical large tank size and high stocking rate for Malawi cichlids make a high

A tank of Malawi cichlids in which the fishes show their health and happiness with upright fins.

flow rate necessary, at least five or six times the total gallonage per hour. Thus a 75-gallon tank would need a filter with at least 400 gph output. Even better would be two filters so that all of your filtration eggs are not in one basket in case of mechanical failure. Since the price of power filters does not normally correlate perfectly with their capacity, two smaller units will probably cost more than a single large one, but, on the other hand, you can probably procure two mid-sized ones for just a little bit more, thereby increasing the overall filtration while providing increased security.

Aeration is important, but the powerheads, lift tubes, or power returns of your filter should be sufficient. I like to use additional powerheads to produce a substantial current, however, not only for a little more biotype authenticity (the waves on Lake Malawi can top twelve feet!) but also because the fish seem to enjoy it. Even small fry are able to hold their own against a powerful flow, and my mbuna often seek out the full current and "surf" it, swimming just enough to stay put against it. In fact, I recently set up a mbuna aquarium and when the fish were put into the tank, they all oriented themselves into the flow from the powerhead and stayed there quite a while, as if to refresh themselves after their trip. It was quite a sight, forty cichlids in formation, swimming in place.

OUTFITTING THE TANK
Forty cichlids in one tank? Well, they were young, and the tank was large, but isn't that a bit

The shores of Lake Malawi often look more like an ocean shore with huge waves which stir up the water and oxygenate the water surrounding the rocks.

A beautiful habitat tank of Malawi cichlid where the tank is supposed to imitate nature.

much? No, because, paradoxically, the best way to minimize mbuna aggression is to crowd them. This is because they have evolved to fit a crowded, bustling world. They are territorial, with breeding males being exceedingly territorial, and with even some females defending foraging territories. The competition for hiding spots, breeding caves, and food is intense, and there is never a dull moment among the rockfish of a Malawi reef.

In the natural environment, their fierceness and belligerence ensure that they will not lose out on their share of resources, and the extremely high population density ensures that there are plenty of targets for this aggression, which is diffused out among the thousands of fishes in the reef dynamic. Captive inside four glass walls, however, it is very easy for the dominant fish to bully a few other individuals, often to death. What you want is to strike a balance between maximum tank carrying capacity and sufficient distraction to diffuse the aggression.

A large tank, properly outfitted, with several dissimilar species of mbuna, and more females than males, will attain a relatively peaceful balance in most cases. What is properly outfitted? In a word: rocks. Rocks, rocks, and more rocks. Now the "rocks" don't have to be rocks; that is, they simply need to be solid objects that afford the fish nooks, crannies, and caves. A big jumble of rocks serves excellently and matches the natural biotype the best, but you can use interlocking plastic rocks, various lengths and diameters of PVC pipe, or even the numerous ceramic or plastic castles

The mbuna graze on the algae which cover the rocks in their natural habitat in Lake Malawi. The rocks are also used as hideaways for many small crustaceans like shrimps and crabs in the same habitat. This grazing characteristic follows the mbuna into your home aquarium and they can be counted on to keep your rocks clear of all algal growth. As a matter of fact, if you don't have enough algae for them to eat, you should supplement their diet with vegetable matter.

Photo by Bernd Kilian.

and ruins so popular today. Your fishes won't really care as long as they have plenty of hiding spots and territorial markers.

If you look at underwater photos of Lake Malawi you'll soon see that we're not talking here about a rock of two on a bed of gravel. You need to pile them in. This raises the concern of the weight of all those rocks, and this hazard is magnified by the fishes' propensity to dig and undermine the rocks, causing avalanches that lead to disastrous damage to the tank glass. Two ways to minimize this danger are to use lightweight artificial (usually plastic) rocks, or to use naturally lighter real rocks such as the

spongy volcanic or tufa rocks. You can epoxy or silicone the rocks together for even more stable formations.

OTHER EQUIPMENT

Your mbuna tank will require all the usual equipment and paraphernalia of a tropical fish aquarium, such as heaters, thermometers, gravel tubes for vacuuming the gravel, and ammonia, pH, and hardness test kits, etc.

As for lighting, since these cichlids are impossibly hard on plants, because of both their digging and their vegetarian tendencies, most mbuna hobbyists forget plants and

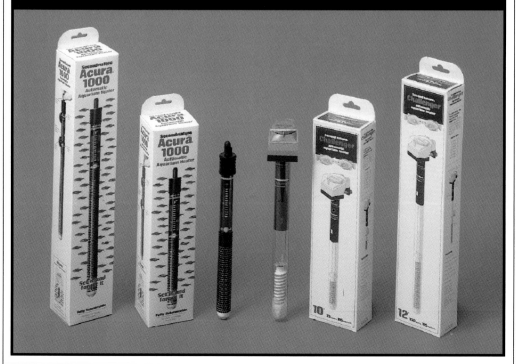

Aquarium heaters, available in both submersible (at left in photo below) and non-submersible models, are made in a range of sizes and wattages to service tanks of different size. They also vary the extent to which they include extra convenience and safety features such as indicator lights, temperature setting indicators. Photo courtesy of Tetra/Second Nature.

simply use regular aquarium fluorescent lighting. If you leave the lights on twelve hours a day you should get some algal growth, and the fishes will happily eliminate it.

A heavy, secure top is a must, since these fishes can and do jump. Besides, it cuts down on evaporation and keeps dust, pets, and half-eaten peanut butter sandwiches out of the aquarium. If you do not use a standard full hood, a piece of glass or acrylic plastic will do. The latter is preferable, for safety's sake as well as for the ease with which you can drill holes for filters, heaters, etc.But with either material, make sure the edges are finished to prevent nasty wounds to passersby or to you while you are servicing the tank.

MAINTENANCE PROCEDURES

Regular aquarium maintenance that you are used to performing is as important in your mbuna tank as in any other aquarium, including checking temperature and pH, testing for ammonia, nitrites, and nitrates, cleaning filters, scraping the front glass (leave the algae everywhere else for the fish to graze on), vacuuming the substrate, and making partial water changes. The last, water changes, are particularly important, for two reasons. First, since you are trying to duplicate one of the cleanest freshwater habitats on the planet, constant water changes are mandatory. Secondly, the high pH of Rift lake tanks, like the high pH of marine systems,

means that ammonia is more a danger than in water of lower pH.

We aquarists tend to be sloppy chemists, and we use the words "ammonium" and "ammonia" loosely. The "ammonia" test kits we use normally test the total of both. The proportion of ammonia to ammonium in a sample differs, depending on pH. You don't have to worry about chemical equilibria; just remember that at a given pH reading the ammonia is more deadly when the pH is higher. To preserve high-quality high pH conditions, you must keep up with the water changes. Many mbuna hobbyists perform and recommend a *minimum* of weekly 50% water changes.

This procedure, easily made part of your regular, habitual tank maintenance, will do more for the health, well-being, and longevity of your pets than a medicine chest full of chemicals or a king's ransom in high-tech equipment, and it will save your fishes much suffering and needless death.

A WORD ABOUT SALT

Many hobbyists routinely add salt, sodium chloride, to their tanks, especially Rift lake tanks. In fact, I may be unusual in not doing so. There are several reasons I do not, but the reasoning is quite straightforward. First of all, I have rock-hard, unchlorinated well water, with a pH of 8 or higher, that I can use right from the tap, and I use soluble substrates and rocks in all my Rift lake tanks to raise and maintain the pH and hardness. This is a major consideration, since I am very fortunate in having this water supply.

Second, while salt does "harden" water in that it adds ions, salt water does not closely duplicate the chemical composition of Lake Malawi's water. Now aquarium fish usually demonstrate great adaptability, and despite all their schooling, none seem to possess a degree in analytical chemistry, so adding salt might be all it takes to make your mbuna feel right at home. I simply prefer to keep as many factors as possible as natural as possible.

And, perhaps most importantly, with about 2000 gallons of water in my aquaria, I need to maximize the ease of water changes. If I add nothing to the water and change it constantly, so that there is little variation between the water in my pipes and the water in my tanks, I know I will keep up with the changes a lot more faithfully than if each water change meant testing, determining how much and which chemicals to add, and retesting to see if I got it right. Salt only complicates matters, since there's no aquarium test kit for sodium chloride, and hydrometers don't discriminate which chemicals are contributing to a sample's density. Without salt I need only an occasional pH and hardness check to make sure

nothing is going awry with my tanks' chemistry. Thus the only tanks to which I routinely add salt are my brackish species tanks; my African cichlid tanks would only receive salt as a medication, if necessary.

There is nothing *wrong* with using salt, and if your water is far from ideal, it can help, but probably buffered Malawi mixtures are a better choice. Mbuna are nothing if they are not hardy little rascals, however, and it often seems that you have to work at it to do them any real harm. Hobbyists often ignore this and take credit for their pets' hardiness, attributing it to their wonderful care. Thus those who salt their mbuna tanks swear by it, citing the lack of salt as a reason for the failure of someone who does not use it to have his mbuna thrive. Those, like me, who forgo the salt and have wonderful success do not quite understand the salt-lovers' fervor for the chemical, but that's no reason to assume someone's lack of success is caused by using salt. Make an informed choice about salt, based on your personal feelings and the chemistry of your water supply, but be open to changing your mind and either adding or avoiding salt if you have less than the success you hoped for in keeping these fishes.

This is a nice sterile Malawi cichlid tank but the fish won't thrive without some algae or vegetation to graze on.

FEEDING MBUNA

While some Malawi cichlids, like *Nimbochromis,* are almost exclusively piscivorous and others, like *Labeotropheus,* are almost exclusively herbivorous, all mbuna benefit from a varied diet. A very good compromise for a mixed species collection is a rotation among:

1. *A high-quality flake or pellet food with a large proportion of vegetable matter.* Pellets are particularly efficient, since there is little waste, and they are available in a gradation of pellet sizes. One trick I use is to feed pellets of various sizes, with the largest just small enough to fit into the largest fishes' mouths. I feed these big pellets first, and while the big guys are working to get one of them down, I throw in the rest, and the littler fish can scoop up the smaller pellets. All the leading brands have excellent formulations especially made for cichlids. Many have high vegetable content, but herbivorous species still need some added green food. A little crushed flake food should be given when you have babies hiding in the rocks.

2. *Live or frozen or freeze-dried invertebrates.* Mbuna, like any other fishes, will relish live food, but they hardly require this temptation to feed. Frozen and freeze-dried daphnia, brine shrimp, and bloodworms are all excellent. Many hobbyists feel that tubifex worms contribute to

"Malawi bloat," a bacterial and/or parasitic infection of the digestive tract. My fishes have never been as fond of them as they are of brine shrimp and bloodworms, so I don't feed tubifex, but if you wish to, use pathogen-free freeze-dried tubifex, which is available.

3. *Vegetables.* All mbuna benefit from vegetable foods, and the strictly herbivorous ones require them as the main portion of their diet. Fed on a high-protein, meat-based diet, these species show spectacular growth, but they will not have normal health. The algae that grow in their tank should only be considered as a snack, however, and you must give them a lot of additional vegetables, with algae being the first choice. You can supply this ideal natural food by placing a container of water in direct sunlight, with plastic plants or rocks in it. When the rocks/plants are coated with algae, put them into the mbuna tank and replace them with ones grazed clean. Algae can also be supplied in the form of *Spirulina* flakes, pellets, or wafers, which the fish will eagerly consume. You can also feed vegetables such as blanched zucchini or squashed peas.

Mbuna are greedy eaters, and aquarium specimens typically outsize their wild cousins by quite a bit. As for most of us accustomed to the easy life, however, this also leads to fatty degeneration of the organs.

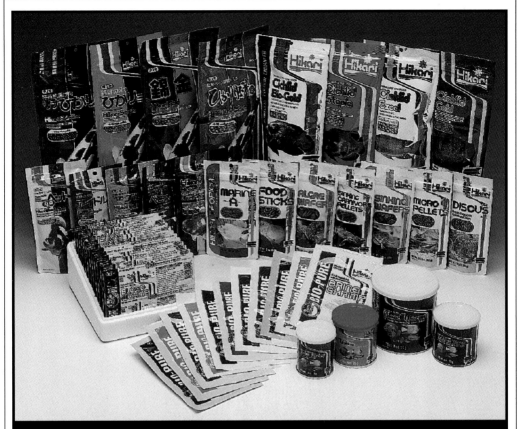

Feeding mbuna is simple because pet shops carry high quality foods which contain the ingredients necessary for Malawi cichlid welfare. This is especially true of pellets. Photo courtesy of Hikari.

Malawi cichlids learn very quickly to beg whenever they see their keeper, and they can be quite persuasive, with forlorn fishy faces pressed up against the front glass. Mine literally jump out of the water when I lift the glass top, vying for first place on the dinner line. For their own good, you must restrict their feeding. Overfeeding leads not only to excessive and unnatural growth but also to problems in maintaining pristine water quality. Remember also that all those rocks provide numerous traps for uneaten food that will decay, destroy your water chemistry, and kill your fishes.

Once they are mature, your fishes can subsist on five or six feedings per week, but a few more than that won't hurt. A day of fasting every now and then is also beneficial. You are undoubtedly used to feeding your breeders heavily to bring them into spawning condition, but lean mbuna are much more likely to produce for you. This is related to the complex social dynamics of Lake Malawi, where competition and defense of breeding and feeding territories keep the fish too busy to gorge themselves. You can feed them daily, but keep 'em hungry!

BREEDING MBUNA

For most hobbyists, breeding their fishes is part of the enjoyment. It indicates that the hobbyist is giving the fishes excellent care, and there is a great satisfaction in looking at a tank full of young fish you raised yourself. Fortunately in this regard, most Rift lake cichlids are willing breeders in captivity, provided a few simple needs are met.

These needs are hardly more than regular maintenance — proper water conditions, good food, adequate space, frequent water changes. Responsible breeders of mbuna are careful to prevent hybridization. While some Malawi hybrids are quite attractive, there is much to lose if hybridization is permitted, and since there certainly is no shortage of variety in color and patterns among the mbuna, there is no motivation to crossbreed species in order to create new morphs. The experience with American livebearers has taught us how impoverished the hobby can become when breeders' lack of foresight permits natural populations to become extinct without adequately pure captive bloodlines.

There are two ways to prevent hybridization. The most sure, of course, is to keep only one species in each breeding tank and to save no fry from your mixed tanks. A second method, which is fairly reliable, is to mix only very different mbuna in a breeding tank, say, one *Labeotropheus* species, one nonsimilarly colored *Pseudotropheus* species, and one *Labidochromis* species. While this does not guarantee that no hybridization will occur, as long as there are members of each sex of each species you should be all right. Make sure that none of the species look like each other, especially in the *females*. Since mbuna fry tend to look like their mothers, if a female releases fry unlike herself it's a tip-off to hybridization.

Remember that it is natural for many mbuna populations to have only a few breeding males, with the rest of the males not defending territories. In the unnatural environment of the aquarium, this can lead to one or two males becoming hyperdominant over all the males of all the species in the tank and breeding with females of any species. You should be alert to this possibility; better still, work to avoid it.

Because mbuna are territorial, aggressive, and polygamous, and because the males are more likely to fight among themselves, the best setup is a large tank housing one or two males and several females of a given species. An abundance of rocks, caves, and other hiding places is necessary not only for the protection and security they provide the fish, but because they will spawn in a cave around which the male defines and defends his territory.

Depending on species, males may dig a spawning cave, construct a spawning pit nest, or even build a "sand castle" with a surface concavity in which spawning takes place.

Spawning is usually quite secretive, especially among the cave spawners, and many people have spawn after spawn without ever observing one. In some species the male passes over the eggs and fertilizes them before the mother picks them up in her mouth; in others, she picks up the eggs and then picks at the male's anal fin, apparently stimulating him to release the milt, which is taken up into the female's mouth, where the eggs are fertilized. Once thought to be necessary for this process, the dummy "egg spots" common on many of these fishes may in fact be misnamed. It is possible that their presence increases the likelihood of a female's having her eggs fertilized, but they are not necessary, and more research is needed to understand the behavioral role of these markings.

After retrieving the eggs, the female usually goes into semi-retirement, lingering in and around the rocks. She will often swim eagerly up at feeding time, but just as often she will either refuse food or just nibble at it. There is some evidence that in certain species the fry feed on food particles that the mother brings into her mouth, producing

different-sized fry upon release. This would certainly extend many of the advantages of mouthbrooding in releasing precocial young. In any case, the female will get thinner and thinner, while her gullet gets bulgier, until finally the emaciated female delivers her brood.

The average incubation/ brooding time is a couple of weeks. During that time you should remove the holding female to a private tank, with lots of hiding places. The females of some species will release the young and then either ignore or try to eat them. Others continue to care for their brood, taking them back into their mouth at any sign of threat. Once it is clear that the young are out for good, you should remove the female. After some serious eating, she will be ready to ripen another clutch of eggs and repeat the process.

Since they are so large on release, all mbuna fry can be started immediately on baby brine shrimp, plus tiny pellets, powdered flake food, microworms, etc. Mbuna are not the fastest-growing cichlids, but they will grow steadily if fed frequently and if you are vigilant with your water changes.

Like many other cichlids, mbuna will breed at a very young age and size. It is normal for fish less than half their adult size to spawn successfully, though, obviously, the clutch size is comparably smaller. This means that your babies will color up (in the dichromatic types) and begin spawning long before they are full size. In fact, the only cichlid I've had breed at a younger age is that old favorite, the convict cichlid, *"Cichlasoma" nigrofasciatum*. I have had a successful spawning of these fish from virtual fry!

COMMUNITY SPAWNINGS

A healthy Malawi community tank is normally also a breeding tank. While most fry will be eaten, the little imps are biologically programmed to survive in a high population, high predation environment. You will undoubtedly see tiny heads poking out from cracks and crevices in the rockpile, and at feeding time a few stalwart tinies will zip out, grab a bite, and zip back in.

In my first mbuna tank I had some rusty cichlids, *Iodotropheus sprengerae*, which I soon had to remove to prevent these milder-mannered fish from being killed by the *Pseudotropheus*. A couple of weeks later I noticed a very small dark fish lurking around the rocks. It was a baby rusty who had somehow survived in that tank full of vicious predators. It was impossible to net, since it never came out of the rocks, so by definition it was too large to be eaten before I could remove it, and even then its life couldn't have been very easy. I can't help but wonder what it thought when I caught it and placed it in a tank of just rusties — out of the frying pan and into paradise!

The female *P. microstoma* picks up the eggs as soon as she thinks the male has fertilized them. Fertilization also takes place in her mouth!

A male *Pseudotropheus microstoma* developing his breeding colors.

Photo by Adam Kapralski.

Having cleaned a spot on the rock, the male flutters his vibrant anal fin egg spot to entice the female to join him.

Photo by Adam Kapralski.

Photo by Adam Kapralski.

Photo by Adam Kapralski.

This remarkable photo shows the female trying to pick up the egg spot on the male's anal fin. Doing this she gets sperm into her mouth. Notice how the egg spot on the male's anal fin looks so much like an egg!

It is clear to see why the female picks up her own eggs so quickly. A predatory *Nimbochromis* has come to eat whatever eggs he can!

Photo by Adam Kapralski.

Photo by Adam Kapralski.

Pseudotropheus tropheops begin their spawning ritual with the female pecking at the male's anal fin.

Newly hatched eggs from the mouth of *Pseudotropheus aurora*.

Photo by Adam Kapralski.

KEEPING MBUNA HEALTHY

Malawi cichlids are generally hardy, easy to keep, and very healthy fish. Kept in optimum conditions, they should stay that way. Of course, accidents happen; the power can go out, chilling the tank and damaging the biofilter cultures, or a territorial dispute can leave a fish with missing scales, torn fins, or even open wounds.

The first response to any such situation should be to minimize all the stress you can. Get the temperature back up, borrow a biofilter from an established tank, step up your water changes, or get the injured fish out into a hospital tank. A reputable dealer will have medications to treat any problems such as ich, fungus, or wounds, plus advice as to what is needed.

Wherever possible, the afflicted fish should be treated in a separate hospital tank, for three reasons. First, it saves the sick fish the added stress of the aggression and competition of the regular tank. Second, it prevents any medications used from killing off your biofilter, and third, it saves you money, since you can treat a small tank instead of all those gallons in the big tank.

QUARANTINE!

That hopefully rarely used hospital tank can actually serve an even more important service in maintaining the health of your collection — by being used as a quarantine tank for new purchases. Quarantining new arrivals is always the only sensible way to add to your stock of fish, but in the case of your Malawi tank, it is especially vital.

Mbuna are not cheap, and there is often some loss of life as a group of these cichlids matures into a stable community. In other words, it takes time and money to establish a tank of Malawi cichlids. Just one infected new fish added to the tank can wipe out all of your investment.

Besides protecting your collection from any diseases or parasites the new fish might be harboring, quarantining protects the new fish from the intense stress that greets any new arrival to a mbuna tank and lets it recover from its transport ordeal before having to face the home crowd on their own turf. When quarantine is up, rearrange the furnishings of your tank before adding the new fish; this will break up all the territories and prevent the newcomer from having to take on everyone at once.

You should quarantine new arrivals for a minimum of two weeks, preferably three or four. If you obtain wild-caught stock, an even longer quarantine period is recommended, since these fishes often bring the greatest risk of infection.

A magnificent female *Pseudotropheus tropheops*. Don't try to breed Malawi cichlids unless they are in good condition like this one collected in Lake Malawi.

Photo by Dr. Herbert R. Axelrod.

In connection with feeding tubifex we've already considered the one malady mbuna seem susceptible to: Malawi bloat. Many mbuna breeders routinely feed new arrivals with a medicated feed to control the protozoan considered responsible for this condition and to prevent losses among both the new fish and the rest of the fish, which would otherwise be exposed when the new ones leave quarantine, since they can carry the disease asymptomatically until they encounter some stress.

The quarantine/hospital tank does not have to be very large unless you routinely add large groups of new fish. A bare bottom is always a good idea, making cleanups easier. You must have caves, of course, but they can be simple lengths of appropriate-diameter PVC pipe, which are easy to move, remove, and clean. For filtration, keep a sponge filter going all the time in an established tank, and pop it into the quarantine tank when you need it for an instantly mature biofilter. Of course, you'll need a heater and a tight cover, but that completes the necessary components of an efficient and vitally effective quarantine/hospital tank.

With mbuna, an ounce of prevention will balance several pounds of cure, and with proper care your pets should enjoy years and years of good health, giving you all those years of enjoyment.